What's Inside?

by Mary Holland

Praying Mantis Egg Case

Insects lay eggs in many colors, shapes, and sizes.

Praying mantises lay lots of eggs in the fall and surround them with foam that comes out of the tip of their abdomen. When the foam hardens, it forms a case that protects the eggs all winter. In the spring the eggs hatch and hundreds of tiny praying mantises crawl out of the case.

praying mantis

Scat
black bear scat

Another name for animal poop is scat. Every animal's scat looks a little different. You can become an animal detective and tell what animal has pooped by the shape and size of its scat. Animal detectives can also tell what an animal has been eating by what's in the scat.

A black bear pooped after visiting a bird feeder. This black-capped chickadee is eating the bird seed out of the black bear's scat.

Bird of Prey Pellet
barred owl pellet

Owls, hawks, eagles, and falcons have trouble digesting the claws, teeth, bones, and fur of the animals they eat (prey).

After they swallow their prey, these hard-to-digest parts stay in their gizzard, a muscular part of their stomach. They are packed tightly together into a pellet and coughed up hours after eating.

By studying the bones and skulls inside a pellet you can tell what animals a bird has eaten.

barred owl coughing up a pellet

bones and teeth found in pellets

Spider Egg Case
black-and-yellow argiope egg sac

Spiders spin several different kinds of silk. Some silk is sticky and is used to catch insects in webs. The silk sac that spiders spin to hold their eggs is not sticky, but it is very tough and waterproof.

Some spiders, like the black-and-yellow argiope, lay eggs in the summer and they hatch in the fall. The young spiders spend the winter inside the egg sac surrounded by very soft silk.

Gall
goldenrod ball gall

The round ball you see on this goldenrod plant is called a gall. A special kind of fly lays an egg on the plant in the spring. After the egg hatches the young fly, or larva, chews into the stem of the goldenrod and the goldenrod swells up around it forming a gall. The fly larva lives inside the gall all winter and eats the walls of the gall for food. When the larva turns into an adult in the spring, it crawls out of a tunnel it chewed when it was a larva and flies away.

gall fly larva

Caddisfly Case
larva and case

Young insects called caddisflies live in the water and build little houses that they carry around with them everywhere they go. Some caddisflies build their houses out of pebbles, some out of twigs, and some out of bits of leaves. After they grow into adults, caddisflies leave their houses and fly away.

Sometimes you can see caddisfly houses moving along the bottom of a pond. Look closely and you may see the caddisfly's head and legs come out when it moves.

Spittlebug Spit
"spit"
spittlebug nymph

Have you ever seen bubbles on the stem of a plant? Inside these bubbles are young insects called spittlebugs. They hang upside down and drink sap from the plant. Bubbles come out of the tip of the spittlebug's abdomen and fall down around it. They hide it from predators and keep the spittlebug from drying out.

When they grow up, spittlebugs are called froghoppers because they can jump a long way, like frogs. They can jump 100 times their own length! If you were a froghopper, you could jump over three school buses end to end.

froghopper emerging from nymphal skin

froghopper

Pileated Woodpecker Nest Hole

Can you find a hole in the tree in this picture? Who do you think made it?

Woodpeckers drill holes in trees with their beak to find insects to eat and to make nests.

When it's time to nest, pileated woodpeckers use their long, sharp bill to hollow out a hole, or cavity, and then lay their eggs inside it. The eggs hatch and the young woodpeckers live in the cavity, protected from the wind and the rain and predators. Their parents bring food to them until they are all grown up and can fly.

Next year the young woodpeckers will make their own nest holes.

Bald-Faced Hornet Nest

Some insects have large families that live together in groups called colonies. In a bald-faced hornet colony there is one queen, a few males and lots of female workers. The colony builds itself a nest to live in with paper they make. Worker hornets chew wood into a kind of paste that turns into paper when it dries. Each different color of paper is made from a different kind of wood. Inside several walls of paper are rows of tiny cups, or cells, into which the queen lays eggs that hatch into larvae. Workers then feed the larvae until they become adult hornets.

A hornet nest is lived in for only one summer—all the hornets except for young, fertilized queens die in the fall. The queens spend the winter in a protected spot such as a rotting log and start new nests in the spring.

inside of nest

bald-faced hornet

Drey

While you might think this is a bird's nest, this is an eastern gray squirrel's nest called a drey.

Gray squirrel dreys are usually made of sticks and leaves high in a tree. Squirrels line the drey with grass, moss, leaves, and shredded bark for warmth for their young. Usually there is one entrance hole, facing the trunk to keep rain out.

A drey is usually inhabited by one squirrel, but two are known to share a single drey in order to keep warm in the winter.

Gray squirrels give birth in late winter and again in the summer. A more protective tree cavity may also serve as a nursery in the winter.

The average drey is only used for a year or two.

eastern gray squirrel

Cocoon
cecropia caterpillar
cecropia moth cocoon

In the spring, a moth begins life as an egg which hatches into a caterpillar. The caterpillar eats and eats and eats in the fall and spins a cocoon out of silk. Inside the cocoon the caterpillar turns into a pupa. The following summer the pupa turns into an adult moth with wings, crawls out of the cocoon and flies away.

dissection of cecropia cocoon

cecropia moths

inside of lodge
Beaver Lodge

Beavers make their own homes, called lodges, by cutting down trees and making a big pile of mud and sticks in a pond. Then they chew one or two rooms inside the lodge where they will live.

Not many people have seen the inside of a beaver lodge as the entrances to it are under water. It is very dark and damp inside a beaver lodge. There is a main room where the beavers dry off and eat and groom themselves and each other, and another smaller room where they sleep. Four to eight beavers usually live inside a lodge.

Black Bear Den

This young boy is exploring the inside of a black bear's den.

Can you think of something you'd like to look inside of? A rotting log? An old bird nest? What can you find to explore?

For Creative Minds

Organ Pipe Mud Dauber Wasp Nests

organ pipe mud dauber wasp cells

Many wasps and bees build nests in which they raise their young. Some make their nests out of wax, some out of paper they make, and some use mud to build their nest. Organ pipe mud dauber wasps collect mud and build mud cells to hold their eggs. Several mud cells side-by-side look like the pipes of an organ, which is how this insect got its name.

You might be surprised if you could see inside an organ pipe mud dauber cell! Each cell contains one wasp egg as well as food for the young wasp when it hatches from the egg.

Young organ pipe mud dauber wasps like to eat spiders. The adult female wasp locates and then stings several spiders. The spiders are alive but can't move after being stung. The wasp carries them back to her newest cell and stuffs them into it. Then she lays an egg on top of the spiders and seals up the cell with mud. When the wasp egg hatches, the young wasp will eat the spiders. Because the spiders are still alive, they don't rot. When the wasp is all grown up, it chews its way out of the mud cell and flies away.

wasp delivering ball of mud and building cell

wasp returning with spider and putting it in cell

wasp cell stuffed with spiders and one wasp egg

Match The Animal With Its Scat

Different animals produce scat with different shapes and sizes. Once you identify whose scat you found, you can find out more about what that animal eats by looking at what is in its scat. Among other things, you might find feathers, seeds, bones, grass, hair, and insects. Can you match the picture of the animal and what it eats with its scat?

black bear –
apple

coyote –
white-tailed deer

beaver –
tree bark

raccoon –
wild strawberries

river otter –
fish

Answers: black bear-5; coyote-4; beaver-1; raccoon-2; river otter-3

Galls

Galls are abnormal growths that form on plants. Sometimes a plant reacts to an insect laying an egg on or in it and grows a bump around it. These bumps, or galls, come in different sizes and colors and shapes, depending on the type of insect that lays the egg and the type of plant it lays it on. Each insect has a specific host plant and a specific looking gall. There can be one or several insects living inside a gall. The gall serves as a shelter and sometimes as a source of food for the insects while they are growing.

Here is a close look at some galls and the insects that live inside them.

Now that you know what to look for, see if you can find galls when you are outside.

blackberry knot gall with wasp larvae

coxcomb elm gall with aphids

pignut hickory gall with aphids

red pouch gall (on staghorn sumac) with aphids

goldenrod spindle gall with moth larvae

A Bird Nest With A Roof!

A bird's nest is usually used to raise one family of birds. After most young birds have left their nest, neither they nor their parents ever return. Other animals make use of the materials that the nest was made out of and sometimes make use of the nest itself.

Deer mice and white-footed mice sometimes make a roof out of milkweed or cattail fluff over an abandoned bird nest in the fall and use the nest as a snug winter home.

This book is dedicated to Otis, Lily Piper, Leo, and children everywhere whose innate curiosity makes them look inside of and explore everything they encounter. Gratitude to Jody Crosby who came up with the concept about which this book was written—MH

Thanks to Kim Hargrave, Education Director at the Denison Pequotsepos Nature Center for reviewing the accuracy of the information in this book.

Library of Congress Cataloging-in-Publication Data

Names: Holland, Mary, 1946- author.
Title: What's inside? / by Mary Holland.
Description: Mt. Pleasant, SC : Arbordale Publishing, LLC, [2024] | Includes bibliographical references.
Identifiers: LCCN 2023052636 (print) | LCCN 2023052637 (ebook) | ISBN 9781643519883 (paperback) | ISBN 9781638170075 (ebook) | ISBN 9781638170266 (pdf) | ISBN 9781638170457 (epub)
Subjects: LCSH: Animals--Miscellanea--Juvenile literature. | Nature--Miscellanea--Juvenile literature. | Natural history--Juvenile literature.
Classification: LCC QL49 .H6843 2024 (print) | LCC QL49 (ebook) | DDC 590.2--dc23/eng/20231204
LC record available at https://lccn.loc.gov/2023052636
LC ebook record available at https://lccn.loc.gov/2023052637

Also available in Spanish: *¿Qué hay dentro?*
Spanish Paperback   9781638173137
Spanish PDF   9781638173199
Spanish ePub3 9781638173229
A dual-language read-along (ISBN 9781638170075) is available online at www.fathomreads.com

Bibliography

Holland, Mary. Naturally Curious: A Photographic Field Guide and Month-by-Month Journey through the Fields, Woods, and Marshes of New England. Second Edition. Trafalgar Square Books. North Pomfret, VT, 2019. Winner, National Outdoor Book Award.

Printed in the US
This product conforms to CPSIA 2008

Arbordale Publishing, LLC
Mt. Pleasant, SC 29464
www.ArbordalePublishing.com